YOUR KNOWLEDGE HAS VALUE

- We will publish your bachelor's and
  master's thesis, essays and papers

- Your own eBook and book -
  sold worldwide in all relevant shops

- Earn money with each sale

Upload your text at www.GRIN.com
and publish for free

Bibliographic information published by the German National Library:

The German National Library lists this publication in the National Bibliography; detailed bibliographic data are available on the Internet at http://dnb.dnb.de .

Imprint:

Copyright © 2016 GRIN Verlag, Open Publishing GmbH
Print and binding: Books on Demand GmbH, Norderstedt Germany
ISBN: 978-3-668-12853-8

This book at GRIN:

http://www.grin.com/en/e-book/313731/managing-the-flow-of-water-from-reservoirs

Elegbeleye Oladipo

Managing The Flow Of Water From Reservoirs

GRIN Publishing

Managing The Flow Of Water From Reservoirs

Elegbeleye Oladipo Ayodamope

Cyprus International University, Lefkosa,

Abstract: *Reservoirs have since been the earliest civilisation, being part of man kind strategy for survival. These water storage facilities have been created since time immemorial mainly by constructing dams accross rivers and are intended both for capturing water that would otherwise proceed down river towards the oceans or lakes as well as for storing the water for as long as possible, till when needed and used. Sediments which is naturaly released from the catchment areas into the reservoir via contributing stream is impeded by the water bodies in the reservoir which are retained by the dams and to a large extent the sediments is permanently trapped there in. This paper talks about managing the flow of reservoirs and regulated river or dam and how to improve its sustainability by balancing the needs of the society, the economy, the environment and the ecosystems. Likewise its environmental goals, its role on a global scale with the laws and regulations guiding it are also discussed in this paper.*

Keywords: dams, ecology, flow , reservoir, water

Contents

1 INTRODUCTION

Water is the most essential of our natural resources, and it is the responsibility of every consumer to ensure that we manage and use it effectively and sustainably. The latest climate change predictions show that pressure on water resources is likely to increase in the future. In light of this, we have to ensure that we continue to maintain and improve sustainable thoughts by balancing the needs of society, the economy and the environment. The natural flow patterns of rivers and creeks can be disrupted by human activity, for example:

People and businesses using river water, dams built to provide drinking water, land being built on, changing the way water runs off surfaces and into rivers and Pumping groundwater found beneath the earth's surface, often between saturated soil and rock crevices – this reduces the amount of water that would otherwise flow into rivers and wetlands.   Reservoir   is   an outdoor storage area whereby water is amassed and preserved in quantity in order that it should be drawn off for use.   Generally, reservoirs are designed by constructing dams across rivers. However, off-channel reservoirs may be provided by redirecting canals and pipelines that carry water from a river to natural or artificial depressions At that point when stream flow is seized in a reservoir, there is reduction in velocity flow and silt is deposited. Therefore, streams that convey much suspended sediment are poor locations for reservoirs. Sedimentation deposits will quickly reduce and diminish storage capacity and greatly shorten the useful life of a reservoir that is small. For larger reservoirs however, siltation constitutes a serious problem because is generally very costly and difficult to remove sediments deposited in reservoirs making the life expectancy of most reservoirs not more than a 100 years. This paper talks about managing the flow of reservoirs and regulated river or dam and how to improve its sustainability, its effect ,its role on a global scale with the laws and regulations guiding it.

Water reservoirs range in different sizes and is multifaceted in nature whereby its complexity ranges from little single-criterion issues to complex multiple-purpose problems. For example a single-purpose reservoir is constructed to achieve only one purpose, such as navigation, flood control, water supply, recreation, irrigation and power generation. Work is still going on in the construction of multiple-purpose reservoirs with the intent of serving at least two major functions. Reservoirs or regulated rivers are two types of close aquatic ecosystems with a typical source or starting point which have a unidirectional association or relationship. One cannot exist without the other and both rise up out of necessity to exploit and take advantage of water resources. Everything that happens inside the reservoir affects the regulated river downstream. In the developed world, social insight especially urban perception take more cognizance of the negative environmental impacts of reservoirs and ignores underestimate and does not value or take cognizance of its positive effects. Because of this, regulated or controlled rivers are frequently subjected to very low flow rates, and mostly appear to be a degraded fluvial ecosystems.

Considering this action , there are some individuals who cast blames on the reservoirs for the tremendous damage sustained by the rivers, and they are actually asking for the demolition and annihilation of dams for different selfish reasons (Poff& Hart,2002; Brufao, 2002), while others value, esteem and regard the involvement and commitment of this kind of artificial ecosystem to modern day welfare due to the control of the available water and energy in time and space because of the existence of reservoirs and regulated rivers (Bergaet al.,2002). Environmental flows are portrayed as the quality and timing of water flows needed to maintain freshwater and estuarine ecosystems and the human well-being that rely on these ecosystems. The past two decades have experienced increased global concern about the need for sustainable water and land management. Water resource availability and use is altered by human growth, economic development, global climate change etc, which leads to increased risk of flow regimes that has been altered, dangers to water quality and water demands which exceeds renewable supply. Water security is mostly achieved with little thought of environmental effects.

Notwithstanding, when they are recognised, the desire for water between humans and the environmental is growing and expanding at an alarming rate. However, the environmental flows concept has kept on developing and growing despite all the difficulties and challenges its facing. Transferability of experience on the field is very limited and in adequate and this is due to the dominance of specific case-study analyses and little or no research on how to govern environmental flows. To implement the environmental flows concept, dialogue needs to be done amongst scientists, researchers, policy-makers, water managers and the indigenes concerning sustainable water utilization. Scientific and Social projection seasoned by earth science have considerably added to the progress of reservoirs and controlled or regulated rivers which encourages and urges solutions rather than theories in the face of sensible and practical water management issues. Reservoirs are superb systems for experimenting and confirming virtually any ecological ideal model (Ward& Stanford, 1984; Straskrabaet al., 1993). It is unquestionably genuine and true that there are badly designed or unnecessarily built reservoirs, and there are doubtlessly taken advantage of, criticizing the situation is inherent, however it may also be viewed as an inherited situation which can and should be improved using environmental management principles.

1.1 Management Of Reservoirs And Their Environmental Goals

Of all the environmental changes and adjustments created by dam development and reservoir operations, the alterations of natural water flow regimes has had the most prevalent and damaging results on the river ecosystems (Poff et al. 1997, Postel and Richter 2003). Related problems of reservoirs is erosion of the stream channel beneath a reservoir when water is released or discharged. Since the sediment or dreg load is deposited within the reservoir, the discharged water has improved transporting capability and channel erosion effects. . The role of water-storage reservoirs, however, is to seize and stop the flow of water in periods of higher flows so as to avoid flooding and allow gradual but steady release of water during periods of lower flows. Reservoirs and regulated rivers started way back in human history primarily to provide and make available water for drinking and for irrigation purposes. The use of reservoirs spread from southern Asia and northern Africa to Europe and the other continents of the world. At that point when stream flow is seized in a reservoir, there is reduction in velocity flow and silt is deposited. Therefore, streams that convey much suspended sediment are poor locations for reservoirs.

Sedimentation deposits will quickly reduce and diminish storage capacity and greatly shorten the useful life of a reservoir that is small. For larger reservoirs however, siltation constitutes a serious problem because is generally very costly and difficult to remove sediments deposited in reservoirs making the life expectancy of most reservoirs not more than a 100 years. The ecological administration of reservoirs must be considered from a global perspective on the conditions that the target area are complete, common and balanced. These objectives are very simple to list but hard to reach. Owing to this, most times when the technical aspects are has been taken care of, some other conditions which include social, economic, and even legal conditions requires more complex, difficult and advanced solutions.

This can be characterized as the excess supply of nutrients in a lake or other body of water, this is due to run-offs from the land and this causes explosive growth of plants and algae in a body of water. A state of hypoxia is reached when organism consume all the oxygen in the body of water resulting to their death.

Eutrophication, the presence of which is more probable are stronger in reservoirs than regulated rivers conditions not minding or taking into considerations their water quality. The two main objectives and goals of eutrophication primarily are to control the level of nutrients that reaches the reservoirs and maximize their processing capacity in the reservoir-regulated river system. The main challenge of environmental sustainable water management is to form, implement and execute a water management program that stores and redirects water for human utilization in a manner that does not influence the ecosystems to degenerate, break down or modify. This journey for parity basically recommends that there is a limit to the quantity of water that can be drawn from a river, and there are also limits in the extent to which the shape of a streams natural flow designs can be changed.

These points of confinement are described by the ecosystem's needs for water. Human control that surpass these limits of confinement will in time compromise the ecological reliability of the affected ecosystems, which brings about the loss of local species and gainful ecosystem products and services for society which are lucrative .With human employment of water and our understanding of ecosystems frequently evolving, the solutions and answers for meeting both ecosystem and human desires will advance overtime likewise. Furthermore, environmentally sustainable water management is an iterative process associated with both human water demands and ecosystem requirements, refined, and modified to meet human and ecosystem needs concurrently once and for all. Water-related sicknesses are mostly linked with the invention of irrigation which is a product of water flow from a typical reservoir such as a dam. Illnesses that are mostly associated directly with irrigation include malaria, and river blindness whose vectors multiply in the irrigation waters. Other irrigation-related health issues include deterioration of water quality and increased population pressure in the area. The reuse of wastewater for irrigation has the ability, depending on the magnitude of treatment and spreading of communicable diseases.

An appraisal of the hydrologic adjustment related with the dam shows major changes in almost all phases of the flow regime, especially in occasions of higher flow. Thurmond Dam also known as Clark hill dam has been very helpful in controlling floods, hydroelectricity, downstream navigation and high-flow pulses. Since the year 1980, floods of different categories (small or large) have been removed, and high-flow pulses larger than 450 m^3/s take place with less frequency]. However, due to water extraction, there has been a 15% reduction annually from 308 to 263 m^3/s on average water discharge (Richter and Thomas, 2007). Water administration is driven and motivated by measured targets, example specified levels of flood protection, generation of hydropower etc. Similarly, ecological goals that are related to water requires quantitative outlining so that they can be integrated and coordinated with other water management objectives (Rogers &Bestbier1997).

Various parts of hydrological variability can affect freshwater biota and ecosystem processes. However in building and developing ecosystem flow, scientists generally cast their gaze on these major parts of flow regimes and these include the wet season and the dry season base flows, high flows that occur normally, severe famine and flood conditions that do not occur annually, rise and fall of flood rates and in addition inter annual variability in every of those parts (King &Louw 1998; Arthington & Zalucki 1998; Trush et al. 2000). The real flow components, segment or statistics used to characterize flow requirements in various parts of the World which vary to some extent, depending upon regional differences in annual hydrologic patterns. Ecosystem flow requirements can be determined as numerical ranges where by the flow element is to be maintained and preserved (Richter et al.). Generally, the more the number of flow characteristics used to describe ecosystem requirements, the higher the probabilities of attaining the desired flow regime. Then again, the flow requirements should be described using only as many traits as essential and needed. It is very possible to recognize a limited number of attributes necessary to explain flow conditions of concern. For instance, even though natural floods are vital in sustaining river ecosystems, their natural variability may not be inhibited in a particular watershed in the absence of dams.

Subsequently there may be no need or reasons to recommend flood flow characteristics unless new dams are projected for future purposes. This may help simplify the evaluation of the ecological suitability of different water management options or alternatives. Primary and Essential considerations should be given to flow features that have been changed by human factor or influence (Rogers &Bestbier 1997; Rogers & Biggs 1999).Population viability depends on a variety of other ecosystem conditions that are not linked to, flow variations, thereby muddling up the relationship between the flow variables and population viability. Evaluation of ecosystem flow requirements should not be restricted to consideration of species needs, notwithstanding, the flow requires individual species to give only a very limited view of the wider range of flows expected or needed to conserve a healthy river ecosystems. Of great importance is assessing the flow conditions and particularly, disturbing effects associated with famine and floods that make-up stream and floodplain ecosystems (Hill et al. 1991; Richter & Richter 2000; Trush et al. 2000). A river's natural flow regime is a basis for determining ecosystem flow requirements. Ecosystem flow should always mimic natural flow attributes to the highest degree possible (Poff et al. 1997, Tharme & King 1998).

It is vital that assumptions and hypotheses concerning flow-biota relationships, other non-flow related variables that affect biota, or influences the flow on other ecosystem conditions like water quality or physical habitat, be made clear when describing the initial estimates of ecosystem flow needed. Creating conceptual ecological models that shows presumed relationships is a superb way of communicating hypotheses (Richter & Richter 2000). Hypotheses should be formulated or conveyed in a manner that permits them to be carefully tested through designed water management procedures. Initial estimates of ecosystem flow needed should be characterised without favouritism to the apparent feasibility of achieving them through near-term alterations in water management. It should be reaffirmed that ecological sustainability should be acknowledged and recognised to be achievable until conclusive evidence suggests otherwise. We have been involved in numerous and various water management conflicts and clashes in which initial perceptions of unfeasibility were conquered through creativity and deeper analysis, with some modification in the political sceneries that made possible what seemed impossible a decade or two earlier.

2.2 The Role Of Water Evaluation In Addressing Global Environmental Flow Requirements (Efrs)

Water scarcity evaluation on a global scale complement regional scrutiny by capturing large-scale improvements in global trends. This evaluations usually ascertain water availabilities and demands at a pixel of 0.5 degree and these values are then aggregated to basins, countries or regions. So far, EFRs have been handled and taken care of in a simplified manner in international water appraisal. The Water Gap model, for instance, has been used to evaluate EFRs by categorizing rivers with respect to the natural variability of their flow regimes. The idea of this approach is that aquatic ecosystems in basins with comparatively stable flow regimes will not be too strong and hence higher minimum flows is required. Water Gap make use of Q90 (the flow amounting to or exceeding 90% of the time) as a low flow ration to denote the mean annual EFRs, but not intra-annual inconsistencies or variability. The H08 model however has been used to measure monthly EFRs likewise categorizing basins according to their hydro- logical conditions (their dryness/wetness and flow variability). The guidelines for classifying basins depends on case studies from dry lands and densely populated regions. Total global EFRs with respect to this study amounts to 30% of total runoff. Studying the water scarcity globally using basin-scale, Hoekstra et al. added a hypothetical environmental flow standard, based on the Sustainability Boundary approach. This assumes that modification beyond 20% in a river's natural flow regime elevates the risk of moderate to major changes to ecosystem services and health. Majority of these approaches and methodologies are practical and not based on any ecological theory or informed analyses and evaluations, this is due to the absence of globally consistent and reliable information, in terms of flow and water use.

3 MAJOR EFFECTS AND CONSEQUENCES OF BUILDING LARGE DAMS AND RESERVOIRS.

NEGATIVE	POSITIVE
Flooding of fertile lands	Regulation of spill over
Population displacement	Supply to cities and industries
Alteration of the landscape	Production of hydro electrical energy
Eutrophication	İrrigation
Alteration of the river's hydrological regime	Recreational use (introduction of exotic species)
Sediment retention	
Barrier effect on fauna Irrigation	

3.1 Laws And Regulations For Reservoirs

All reservoirs that can accommodate at least 25,000 cubic meters or more, that could escape in the event of a dam failure, must be registered with the Environment Agency (EA). These are known as large raised reservoirs. They are large raised structures designed or used for collecting and storing water; these structures can include large lakes.

3.2 How Reservoirs Are Controlled Or Regulated

Once a dam or reservoir is registered or enrolled, the Environmental Agencies will decide whether or not our reservoir is high-risked or not. A reservoir is 'high-risk' if, there is incident of an uncontrolled discharged of water from the reservoir, thereby causing human life to be in danger. If our reservoir is 'high-risk', the full regulations will apply.

1. The reservoir will still need to be registered but
2. It won't be subject to full regulations

It is essential that 'not high-risk' reservoirs are also registered in order that their designation be evaluated if there is any alteration of circumstance, for instance, an improvement in the downstream flow. Till the Environmental Agencies makes a decision on whether or not our large raised reservoir is 'high-risk', it will continue to be fully regulated which means a supervising engineer will be appointed and there will be an inspection every 10 years

4 DESILTING RESERVOIRS

Eutrophication is a characteristics procedure at the early stage of any reservoir where by filling is required and sediments maintenance is no less important for the de-silting of reservoirs, sediment distribution in the interior of the reservoir, together with their size and chemical characteristics, are essential points to consider when planning any action. One option is dredging, the efficiency of which has always been questioned nevertheless some experiences with acceptable results can be seen even for deep reservoirs (Jacobsen, 2003). After extracting the sediment there is the problem of disposal. The ideal would probably be to gradually deposit it downstream, in combination with hydrological- forestry stabilisation actions at the source basin. Normally it is transported to dumping areas with a variety of possibilities, ranging from composting, to the recovery of degraded areas. In any case prior dehydration is usually required (drying, compression...) and sometimes chemical treatment is necessary, either for stabilisation (Murakami et al., 2004) or for agronomic reuse with the option to exploit the accumulated organic phosphorous.

Another option is to completely drain the reservoirs using the bottom outlets. In this way it is possible to remove a considerable volume of sediment downstream. In Spain, a number of large reservoirs have already been emptied and drained for different aims (Santa Ana, Barasona, Alloz, Doiras, Sallente, etc.), which have served to provide empirical data concerning the more significant aspects of this type of action. To date, the best-documented experience is undoubtedly the case of the Barasona reservoir (river Ésera, Huesca), where it was possible to analyse the considerable environmental impact of this type of action, both at the reservoir itself and in the downstream section of the river. A collection of the main monitoring work carried out during the draining of the Barasona Reservoir has been published in Limnetica (1998; vol. 14). It was possible to confirm that the effects of the action were totally reversible in the short to medium term and the positive end results for the aquatic system (reduction of the nutrient load and elimination of exotic fish species in the reservoir, recovery of amphibian populations, improvement of downstream river banks, etc.).

The best option in the prevention and correction of silting up in operational reservoirs, is once again, however, the management of stored water levels, periodic control of the outlets, depth water discharges of the dam and the application of regular floods. Consequently, the maintenance of low reservoir levels, together with the use of bottom outlets at times of natural floods, is the best way to mobilise sediment inside the reservoir towards the dam area in order to evacuate it downstream, as well as maintaining the pipes and drainage systems of the reservoir clean and in working order.

4.1 Identifying Incompatibilities Between Human And Ecosystem Needs

Regions of possible incompatibility between humans and the ecosystem needs must be thoroughly observed and investigated. Within-year assessment shows the precise months or seasons in which ecosystem flow requirements are not likely to be met. Assessments of various years will enable easy understanding and comprehension of the frequency with which ecosystem needs could be damaged and abused. Regions of possible inconsistencies that are not compatible with human and ecosystem needs should be assessed and monitored for every stream reach of concern, as the nature and character of conflict can shift widely from upstream to downstream. Models can be used to discover water management alternatives to recognize and highlight the trivial contrast between choices consequently constraining the scope of the variance (Carver et al. 1996). Statistical evaluation of the variations between human-influenced flow conditions and ecosystem requirements can measure the magnitude of potential conflicts and variances (Richter et al. 1996).

5 CONCLUSION

With the increasing effect of climate change, the water flow has become more of a nuisance than a natural activity due to the increasing sea rise, increasing rainfall and melting glaciers which are two intertwined factors that have a reverberating effect on water flow especially from reservoirs. The environment is adversely affected by the reservoirs especially their water flow which in many instances cause eutrophication, etc. The management of water flow from a reservoir as discussed above is very important as it impacts the economy, the ecosystem and living matters both unicellular, multi-cellular and complex organisms. Scientific inventions such as the application of computer models to investigate water flow and its management cannot be over-emphasized.

6 REFERENCES

Arthington, A. H. (1994). *A holistic approach to water allocation to maintain the environmental values of Australian streams and rivers: A case history.* Mitt.Internat.Verein.Limnol vol. 24, p. 165-177.

Duncan (eds.). p. 213-288. Kluwer Academic Publishers.Netherland.

Encyclopaedia Britannica.

http://www.britannica.com/EBchecked/topic/499101/reservoir

Jacobsen, T. (2003). *Sediment handling technologies: experience from case studies. Hydropower &Dams.* Vol. 6, p. 84-87.

King, J. and D. Louw. (1998). *In stream flow assessments for regulated rivers in South Africa using the Building Block Methodology.* Aquatic Ecosystem Health and Management vol 1, p. 109-124.

Pahl Wost (2013) . *Water Governance and Environmental Flow: managing sustainable water uses,* Curr Opin Environ Sustain (2013) http://dx.doi.org/10.1016/j.cosust.2013.06.009

Poff N.L., J. D Allan, M.B. Bain, J. R. Karr, K. L. Pretegaard, B.D. Richter, R. E. Sparks and J. C. Stromberg. (1997). *The natural flow regime: a paradigm for the conservation of rivers and restoration.* Bioscience Vol. 47, p. 769- 784.

Poff, N. L. & D. D. Hart. (2002). *How Dams Vary and Why It Matters for the Emerging Science of Dam Removal.* Bioscience, 52(8), p. 659-668.

Renata J. Romanowicz , Adam Kiczko &Jarosaw J. Napiórkowski (2010) *Stochastic transfer function model applied to combined reservoir management and flow routing,* Hydrological Sciences Journal, 55:1, p. 27-40 http://dx.doi.org/10.1080/02626660903526029

Rogers, K. and R. Bestbier. 1997. *Development of a Protocol for the Definition of the Desired State Of Riverine Systems in South Africa. Pretoria: South African Wetlands Conservation Programs, Dept. Of Environmental Affairs and Tourism* Vol. 47, p. 769- 784.

Straskraba, M. (1999). *Retention Time as a KeyVariable of Reservoir Limnology. In: Theoretical Reservoir Ecology and its Applications.*

International Institute of Ecology, BrazilianAcademy of Sciences and Backhuys Publishers, p. 385-410.

Straskraba, M., J. G. Tundisi & A. Duncan(1993). *State-of-the-art of reservoir limnology and water quality management. In: Comparative Reservoir Limnology and WaterQuality Management*. p. 385-410.

Tharme, R.E. Environmental *Flow Assessment a Global Perception. Rivers Research and Application*, in Review.

Trush, W. J., S.M. McBain, L.B. Leopold. (2000) *Traits of an alluvial river and their relation to water policy and management.* Proceedings of the National Academy of Sciences 9

Ward, J. V. & J.A. Stanford. (1984). *The regulated stream as testing ground for ecological theory In: Regulated Rivers.* Vol. IV: 3-15.

6.1 Internet Sources

http://naedacf.pbworks.com/f/Ecologically+Sustainable+Water+Mangement.pdf

http://www.esajournals.org/doi/full/10.1890/10510761%282003%29013%5B0206%3AESWMMR%5D2.0.CO%3B2

http://www.academia.edu/1432471/Ecologically_sustainable_water_management_managing_river_flows_for_ecological_integrityhttp://palmerlab.umd.edu/Publications/Pahl-Wostl_etal_2013.pdf

http://www.researchgate.net/publication/250916376_Environmental_Flows_and_Water_Governance_Managing_Sustainable_Water_Use

http://www.sciencedirect.com/science/article/pii/S1877343513001425

YOUR KNOWLEDGE HAS VALUE

- We will publish your bachelor's and
  master's thesis, essays and papers

- Your own eBook and book -
  sold worldwide in all relevant shops

- Earn money with each sale

Upload your text at www.GRIN.com
and publish for free